ISBN 978-1-5279-1258-8
PIBN 10909159

1 MONTH OF FREE READING

at

www.ForgottenBooks.com

By purchasing this book you are eligible for one month membership to ForgottenBooks.com, giving you unlimited access to our entire collection of over 1,000,000 titles via our web site and mobile apps.

To claim your free month visit:

www.forgottenbooks.com/free909159

English
Français
Deutsche
Italiano
Español
Português

www.forgottenbooks.com

Mythology Photography **Fiction**
Fishing Christianity **Art** Cooking
Essays Buddhism Freemasonry
Medicine **Biology** Music **Ancient**
Egypt Evolution Carpentry Physics
Dance Geology **Mathematics** Fitness
Shakespeare **Folklore** Yoga Marketing
Confidence Immortality Biographies
Poetry **Psychology** Witchcraft
Electronics Chemistry History **Law**
Accounting **Philosophy** Anthropology
Alchemy Drama Quantum Mechanics
Atheism Sexual Health **Ancient History**
Entrepreneurship Languages Sport
Paleontology Needlework Islam
Metaphysics Investment Archaeology
Parenting Statistics Criminology
Motivational

Technical and Bibliographic Notes / Notes technique et bibliographiques

The Institute has attempted to obtain the best original copy available for filming. Features of this copy which may be bibliographically unique, which may alter any of the images in the reproduction, or which may significantly change the usual method of filming are checked below.

L'Institut a microfilmé le meilleur exa été possible de se procurer. Les dét plaire qui sont peut-être uniques du p ographique, qui peuvent modifier une ou qui peuvent exiger une modificatio ode normale de filmage sont indiqués

- [✓] Coloured covers /
  Couverture de couleur

- [ ] Covers damaged /
  Couverture endommagée

- [ ] Covers restored and/or laminated /
  Couverture restaurée et/ou pelliculée

- [ ] Cover title missing / Le titre de couverture manque

- [ ] Coloured maps / Cartes géographiques en couleur

- [ ] Coloured ink (i.e. other than blue or black) /
  Encre de couleur (i.e. autre que bleue ou noire)

- [ ] Coloured plates and/or illustrations /
  Planches et/ou illustrations en couleur

- [ ] Bound with other material /
  Relié avec d'autres documents

- [ ] Only edition available /
  Seule édition disponible

- [ ] Tight binding may cause shadows or distortion along interior margin / La reliure serrée peut causer de l'ombre ou de la distorsion le long de la marge intérieure.

- [ ] Blank leaves added during restorations may appear within the text. Whenever possible, these have been omitted from filming / Il se peut que certaines pages blanches ajoutées lors d'une restauration apparaissent dans le texte, mais, lorsque cela était possible, ces pages n'ont pas été filmées.

- [ ] Additional comments /
  Commentaires supplémentaires:

- [ ] Coloured pages / Pages de couleur

- [ ] Pages damaged / Pages endommag

- [ ] Pages restored and/or laminated /
  Pages restaurées et/ou pelliculées

- [✓] Pages discoloured stained or foxed
  Pages décolorées, tachetées ou piqu

- [ ] Pages detached / Pages détachées

- [✓] Showthrough / Transparence

- [ ] Quality of print varies /
  Qualité inégale de l'impression

- [ ] Includes supplementary material /
  Comprend du matériel supplémentair

- [ ] Pages wholly or partially obscure slips, tissues, etc., have been ensure the best possible image / totalement ou partiellement obscu feuillet d'errata, une pelure, etc., on à nouveau de façon à obtenir image possible.

- [ ] Opposing pages with varying col discolourations are filmed twice to best possible image / Les pages ayant des colorations variables ou orations sont filmées deux fois afin meilleur image possible.

1	2	3

1
2
3

MICROCOPY RESOLUTION TEST CHART

(ANSI and ISO TEST CHART No. 2)

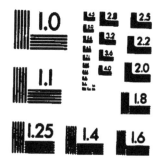

APPLIED IMAGE . Inc

1653 East Main Street
Rochester, New York   14609   USA
(716) 482 - 0300 - Phone
(716) 288 - 5989 - Fax

REPORT

ON THE

QUEEN MINES
SALMO, B.C.

—BY—

ALEXANDER SHARP
MINING ENGINEER

REPORT ON QUEEN MINES

SHEEP CREEK, SALMO, B. C.

Synopsis—Queen Mines

Situated in Sheep Creek Camp, Kootenay District, British Columbia

---

Free Gold Proposition.

Fully one millions dollars produced to date.

$511,761.04 since present owners bought mine, 5½ years ago.

399.26 acres of Mineral Claims.

Queen Vein where ore mined averages at least 9 feet wide.

Developed to a depth of 800 feet from surface.

A 20 stamp mill driven by water power.

Company owns valuable timber for mining purposes and saw mill.

Operating Plant, Buildings, etc., cost value, $100,000.00.

And very valuable water power rights.

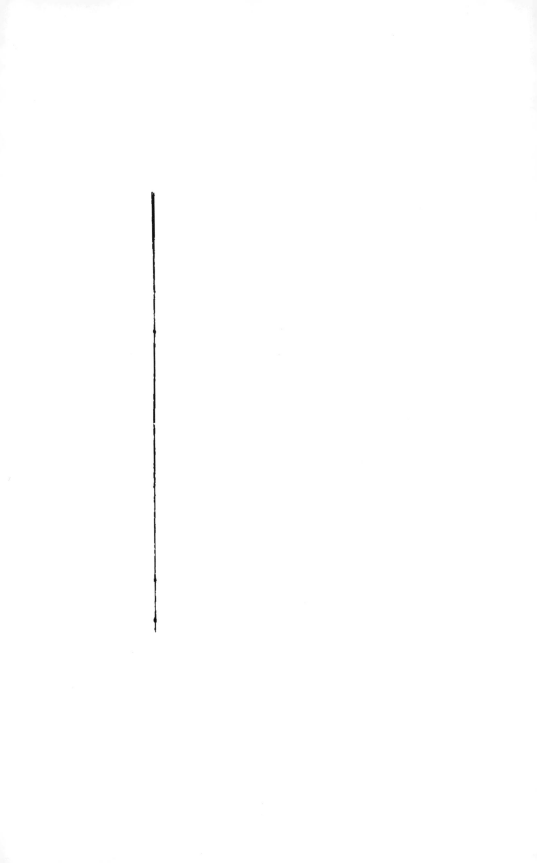

REPORT ON QUEEN MINES

SALMO, B. C.

by

ALEXANDER SHARP
Mining Engineer

(All rights herein reserved to Smith Curtis)

Vancouver, B. C.,
1st December, 1913.

To Smith Curtis, Esq.,
  Vancouver, B. C.

PRINCIPAL DATA ON WHICH THIS REPORT IS BASED

1. R? of topographical and geological survey of surface and underground ...king by myself.

2. Plan of claims and underground workings supplied by E. V. Buckley, Manager Queen Mines.

3. Samples of ore taken by myself and records of ore shipped, supplied by Mr. Buckley.

4. Assays of above named samples by E. W. Widdowson, Chemist, Nelson, B.C.

SITUATION OF PROPERTY

The property is situated at the Junction of Sheep 1 Wolf Creeks, 12 miles East of Salmo, B. C., a town on the Nelson and Fort Shepherd branch of the Great Northern Railway, Sheep Creek Mining Camp, Nelson Mining Division, British Columbia.

TOPOGRAPHY

The topography of the Sheep Creek District is that typical of the Kootenay Mining Country, rugged and mountainous, thickly covered with large timber, much underbrush. In the valleys there is a thick dark red soil. The mountain peaks are bare and more or less rounded, the slopes are covered with drift and wash, while the lower hillsides and valleys are full of large beautiful timbers. Prospecting for mineral is difficult, except on the mountain tops and creeks.

The maximum range in altitude above the Salmon River Valley at Salmo (2200 feet above the sea) is about 3,500 feet. The immediate district around the property is drained by the Wolf and Sheep Creeks, which join together at the property (3200 feet elevation above sea level) forming the main Sheep Creek which empties its waters into Salmon River near Salmo and so flows on into Pead d'Oreille River, a tributary of the Columbia River.

During winter there is considerable snow on the mountains. Temperature seldom falls below zero. The summer bright and cool. Rainfall is light.

EXTENT OF PROPERTY AND TITLE THERETO

The mining property consists of fifteen claims held in fee simple from the Crown, having an aggregate area of 399.26 acres. They are located as shown by Map No. 2, the area of each claim being as under:—

Argyle	29.06	acres, being Lot No.			10155
Wolfe	22.06	"	"	" "	3856
Malwaaz	13.00	"	"	" "	3652
Yellowstone	50.08	"	"	" "	3651
Burlington Fra.	15.50	"	"	" "	1079
Mat	3.17	"	"	" "	3857
Pat	19.41	"	"	" "	5706
Hide Away	28.07	"	"	" "	5625
Queen	34.45	"	"	" "	1076
Alexandra	51.37	"	"	" "	9078
Edward VII.	18.22	"	"	" "	9077
Bullion	50.74	"	"	" "	8325
Niagara	24.77	"	"	" "	1077
Lewiston	31.09	"	"	" "	1078
Placer Fraction	7.73	"	"	" "	9079

WATER AND TIMBER RIGHTS ON THE PROPERTY

The Queen Company own the first rights to (400) four hundred miner's inches of water from Sheep Creek, and to four hundred (400) miner's inches of water from Wolf Creek, and has by a dam reservoired water in a lake near the head of Wolf Creek, having an area of 23 acres. It owns the timber on the claims and the surface rights for mining purposes, and can acquire at any time at $5 per acre from the Government the full surface rights for all purposes including townsite. There is a large amount of timber on the claims available for mining purposes.

GEOLOGICAL FEATURES

The rock underlying the claims is a more or less altered felsitic slaty schist, quartzite, and igneous dykes. The planes of the formation have a strike from S 20° W (Magnetic) to S 45° W, dip from 65° towards North to as much as the perpendicular.

VEINS

There is a number of veins on the property (which are indi... ¹ by lines on Map No. 2) which have a general trend of S. W. and N. E. o.p towards the North from 46° to the perpendicular. The Queen, Yellowstone and Alexandra veins are the principal ones opened out so far. These veins will be fully described under the heading of "Mining and Work Done." There does not appear to have ever been any particular effort made to find veins not clearly exposed on the surface, which is mostly covered by wash, etc., and the chances of other veins being found by systematic explorations on the surface and by underground cross-cut I consider very good

The veins are true fissure in the schist. The schist has been replaced by Quartz, the Quartz forming the veins. Scattered irregularly in small and variable quantities throughout the Quartz are the following metallic minerals: Iron sulphides, galena, and tungstite. Gold in its free state is found in the Quartz and the minerals.

The foot and hanging walls or sides are of quartzite, which contains some mineral, but generally the minerals fade away in the direction of the schist or country rock. The veins appear to be more fully mineralized and richer in precious metal when the wall rocks of quartzite are strong and thick. Everything would indicate that the enrichments occurred from an ascending solution.

HISTORY

The "Queen" and "Yellowstone" were first discovered about fourteen years ago. For three years mining was done intermittently on the "Yellowstone" by lease-holders, who did little development. Thereafter, good surface ore being opened on the "Queen", mining has ever since been confined to it and more systematic mining has been done. The mines so far have produced to 30th Sept., 1913, according to Government returns, $1,030,914 in Gold after deduction of freight and treatment charges; about 53% of this amount was saved by amalgamation at the mine mill; the balance from concentrates, which were shipped to Trail Smelter where $5.40 per ton was deducted for freight and treatment.

MINING AND OTHER WORK, AND RESULTS OBTAINED

The Queen mines are the most important mining property in the Sheep Creek District, and are situated on Wolf Creek, three hundred yards above the forks of Wolf and Sheep Creeks (see Map No. 1). The mines are what is known as a free gold proposition, have extensive development, which consists of three adit tunnels driven on the vein (No. 1, No. 2 and No. 3). No. 3 is the lowest tunnel, its portal is 50 feet above the creek and has been extended in for a distance of 900 feet. All these tunnels are connected by raises or stopes. All the ore from the mine is taken out by No. 3. Three chutes or chimneys of ore were mined from this level, having a length of from 100 to 200 feet. From No. 3 level a double compartment winze or shaft has been sunk to a depth of 500 feet. From this shaft No. 4, No. 5 and No. 6 levels (No. 6 level being 400 feet below No. 3 level), have been run, each for a distance of fully 700 feet, while No. 7 level is at present being opened out from the shaft bottom. The mine has been developed entirely on the vein laterally for a distance of fully 1000 feet, and the bottom of the winze or shaft is 800 feet vertically below the surface. Sketch Map No. 3 illustrates these underground workings. It will thereon be noticed that levels No. 5 and No. 6 have been extended under the portal of No. 3 tunnel and Wolf Creek, towards the "Yellowstone" claim; here another important chute of ore fully 150 feet long, and from 11 feet to 26 feet wide, is being mined, and the present output of the mine is being obtained. It would be difficult to overstate the strength and regularity, at this place, of the vein with its well defined vertical walls and showing its greatest width at the greatest depth stoping has been done in the mine, with ore and values still in the face of the drift. There is great chance of this vein being developed into the Yellowstone property and an important mine being opened out there.

VEIN

The strike of the vein is S 36° W (Magnetic). Its width has varied from 2 to 26 feet wide. These minimum and maximum widths have been for short distances only, the most of the vein showing ore from 7 to 16 feet wide and averaging, I estimate, about 9 or 10 feet. The ore is a coarse grained quartz. The mineralization consists of free gold, iron sulphides carrying gold, galena and some little tungstite, the mineralization averaging about 7%. The hanging and foot walls are strong, hard quartzite, the formation slaty schist. The largest amount of mineral is found in proximity to igneous dyke rock that cuts into the quartz. Above No. 5 level (see Map No. 3) the most of the ore has been stoped out, while under the same level there is much unstoped ground requiring further exploration. The shaft at No. 7 level (lowest point reached) is in quartz and should be opened up in an east and west direction. The Queen vein would appear to be a most important one because of its width and continuity, and the values in gold, produced to date, from $9 to $12 per ton, fully one million dollars gross.

The tonnage mined and milled by present owners during the past five years is 61,427 tons, and recovered $511,761.04 in gold, average recovery $8.33 per ton. From the best estimate I can form from all the information I can gather, the loss in tailings has been 20% to 30%, or $2.00 to $4.00, which would give average gold values in the ore milled of $10.80 to $16.00 per ton. In another mine in the immediate vicinity, equipped with stamps, tube mill and cyanide plant, and a skilled experienced manager, the ores treated are, I am told on excellent authority, showing a gold recovery most of the time about 98%, or a loss of only 20 to 30 cents per ton.

There is no geological reason why, if this vein were explored further, both laterally and in depth, the Queen Mine should not become one of the great mines in the North-west. Hitherto, the management has for some reason been contented to work this mine in a very small way. No doubt this was partly due to the lack of milling capacity; and it must not be forgotten that the mines have been operated for the past twelve years under first one ownership and then another, by an owner or co-owner who never had had any previous knowledge of mining whatever, and it is a remarkable fact that in spite of want of technical knowledge and previous experience they have obtained such successful results, and these, too, with a milling plant not in a high state of efficiency and without tube mill and cyanide plant. No assayer is kept to prevent ore here and there too low in grade from being broken down and milled, thereby lowering the average value of ore treated, and mine map of the workings has not been added to for four years.

During the past history of the mine, mining costs, I believe, averaged $4.11 per ton, milling $1.90 per ton. There is no doubt with a larger output, new additions of latest designed plant, and highly skilled superintendence, these could be materially reduced.

MINE RECORD

Mr. E. V. Buckley, the Manager of the mine, furnished me with figures showing that since his company took hold of the mine five (5) years ago, the mine has produced in gold bullion and concentrates $511,761.04, a very fine record for a mine working in a small way, with only from 20 to 30 men, and unnecessarily losing for want of an up-to-date plant so large values in the tailings. At present the mine is producing about 45 to 50 tons per day; owing to labor troubles the mine has worked short-handed during the whole of this year, and only stoped some 7,000 tons of ore, which was ready to hand, without doing any development.

MILL AND MILLING

The mill is situated near the junction of Wolf and Sheep Creeks, about 300 yards from the mine and connected to the mine by a level tram road, over which the ore is carried in cars, having a capacity of about 1½ tons. The milling plant consists of four batteries of five stamps each, and four Wilfley concentrating tables of the Overstrom-Wilfley type, and a jaw crusher.

The ore is crushed by jaw crusher and batteries to about 50 mesh screen, then passed over the amalgam plates, where about 53% of the recovered value is collected in gold bullion, the remainder is obtained in concentrates, consisting of iron pyrites, galena, and some zinc blende, carrying about 2½ ounces of gold to the ton. The concentrates are shipped by wagon to Salmo, thence by railway to Trail Smelter. The freight and smelter treatment rate is $5.40 per ton. My sampling of the tailings showed a loss at the mill of $1.20 per ton of ore treated, the ore being, I infer from my sampling in the stope, lower than the average grade. but I understand the general loss is from $1.00 to $4.00, and occasionally as high as $6.00 per ton. The mill has a capacity of 60 tons per day.

The power obtained to operate the mill and other machinery at the mine is developed by a Pelton Wheel, under two heads of water from Sheep and Wolf Creeks, of 350 to 480 feet, served by a flume on each creek, about one mile in length, 24 inches by 30 inches. A Rand 10 drill compressor furnishes air for the mine drills, machine and blacksmith shops, hoisting engine and pumps. A smaller compressor is at times brought into use when greater air power is required.

To operate this machinery 250 to 300 horse-power are required, derived from about 300 miner's inches of water of the 800 inches owned by the mine company. There is water enough in the creeks six months of the year to develop from 3,000 to 5,000 horse-power. In the case of any shortage of water during the dry seasons, the 23 acres of lake referred to could be utilized. It will thus be seen there is plenty of power for an increase of the mine and mill output.

The intrinsic value of this water power should not be overlooked. With an ore output of 150 tons per day, 750 horse-power produced from the 800 miner's inches (=22.4 cubic feet per second) would be required. Were this power produced by steam, the cost would be not less than $60 or $70 per year per horse-power, or $45,000 to $50,000, and the timber along the mountain slopes now protecting the water supply would be rapidly used up; or if electric power were brought in from the City of Nelson, a guaranteed contract for a period of years would first have to be entered into to get power delivered at $42.50 per horse-power per year, or a yearly charge for 750 horse-power of over $30,000, or 10%

yearly (to cover 6% interest and 4% sinking fund) on a capitalization of $300,000. This very large cheap power which no other mining property in the vicinity possesses to any such extent, will always make many good adjoining properties subservient to the Queen Reduction Works either for ore treatment at a good profit, or to enable the Queen owners to acquire them when desired or required at a much lower valuation than they would have, had they equal power costs.

YELLOWSTONE MINE

The Yellowstone Mine, owned by the Queen Mines, . . orporated, is situated on the steep hillside 400 feet above the Queen Mine mill. The formation and grade of ore are the same as those of the Queen Mine. It i. .: been developed by two tunnels on the strike of the vein for a distance of 900 feet. The vein is from 3 feet to 8 feet wide. Government returns show the mine has produced gross $124,331 worth of ore. The present company has not operated this mine, owing, I believe, to the mill not being able to handle the ore. I was not able to see this mine, some of the drift and stopes are caved in somewhat, but I understand from a sketch of the mine furnished me, there is considerable ore in the stopes ready for milling. (Map No. 5).

ALEXANDRA CLAIM

The Alexandra vein is also on the Queen property. The amount of snow on the mountains prevented me from seeing this prospect which has been opened out under the present management. A 100 foot drift has been driven in on the vein, and a winze sunk 40 feet. The vein is from two feet to four feet wide and Mr. Buckley informs me, assays as high as $278 in gold obtained.

SAMPLING AND ASSAYING

I made a sampling of the working stope of the Queen Mine and some of the old drifts. Results as follows:

The number of assays are marked on sketch map No. 3, indicating what part of the mine the samples were taken from.

The first ten were from the Buckley stope.

Assays were made for gold only and the value is computed at $20 per ounce.

Buckley Stope	No. 1	$13.20
" "	" 2	7.20
" "	" 3	11.60
" "	" 4	1.20
" "	" 5	1.80
" "	" 6	3.20
" "	" 7	4.00
" "	" 8	3.60
" "	" 9	4.00
" "	" 10	15.00
Extreme bottom of shaft	" 11	.60
Face of No. 6 East drift	" 12	2.80
Cross-cut No. 6 drift	" 13	.40
Pump Station No. 6	" 14	.40
Face No. 3 drift	" 15	Trace

20-STAMP MILL, QUEEN MINE, SHEEP CREEK, B.C.

LOOKING UP SHEEP CREEK FROM QUEEN

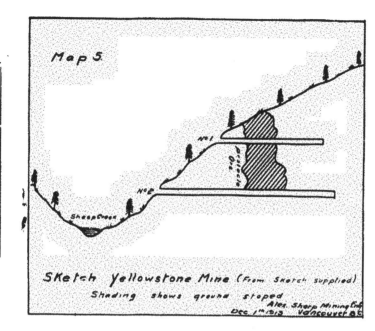

Map 5.

No 1

No 2

Sheep Creek

Sketch Yellowstone Mine (From Sketch supplied)
Shading shows ground stoped
Alex. Sharp Mining Eng.
Dec 1st 1913   Vancouver B.C.

ROM QUEEN MINE AT MOUTH OF WOLF CREEK

The most of these samples were taken from places outside the ore chutes. The ten samples taken from the wide Buckley Stope show that there are good big bodies of pay ore there, but there are parts being mined that careful assaying would show should be left untouched, while the grade of ore milled is being unfairly reduced with corresponding loss of profits.

The other samples were taken in what were known to be more or less barren parts of the vein, so that I might know what kind of ore was left untouched. The assay No. 12 would indicate that values are coming in and that a pay shoot is likely near in East Drift of No. 6 Tunnel.

MINE AND SURFACE IMPROVEMENTS

In addition to the mill plant, compressor, machinery, etc., already described, the mine is well equipped with mine air drills, pumps, hoisting engine and a saw-mill, besides an electric lighting plant.

```
Office Building. . . . . . . . . .32' x 32'
Assay Office. . . . . . . . . . . .12' x 16'
Residence. . . . . . . . . . . . . .16' x 34'
Rooming House. . . . . . . . . .25' x 37' two storeys and a half high
Mill Building. . . . . . . . . . .54' x 64'
Mill Building. . . . . . . . . . .24' x 48'
Store Room
Stables
Air Pipe Lines, etc.
```

The mine equipment plant is worth about ($100,000) One hundred thousand dollars. It must have cost all of that to install it.

SUMMARY

The Queen Mine is in the best free gold belt in British Columbia and in the vicinity of other well-known free gold mines, namely, The Kootenay Bell, Nugget, Mother Lode, Vancouver, etc. It embraces   level of the creek and is well situated for deep tunnel mining. There are fc known veins on the property and doubtless the Kootenay Bell veins run into the Yellowstone. I made an examination of the Kootenay Bell mine three years ago and came to that conclusion. The Vancouver vein may extend into the Alexandra claim. It is my opinion that further development on the Queen Mine, laterally and in depth, will open up new ore bodies on the proven veins and if these veins persist in their present known direction to the boundaries of the mine property, the distance yet to be explored longitudinally will be nearly four times the distance already opened up. Studying the mine sectionally, I came to that conclusion. The four chutes of ore may be traced to run into each other. The west chute is by far the strongest of the four and carries pay ore continuously from the outcrop to between No 6 and No. 7 levels, while the east or Buckley chute is strong, wide and well defined with pay ore on No. 6 floor. A winze sunk in the floor (see X Map 3) here and drifts run from it on the vein would, if the ore persists in depth, and the vein in its present width of from 11 to 26 feet, as they are likely to do, very cheaply and quickly open up very large reserves of ore. An upraise (see XY Map 3) through the ore to the surface on the east side of Wolf Creek would open up more ore, give ventilation, and the installation of the ore hoist here would greatly shorten the route of ore from stope to mill. On No. 6 level, there is some slight igneous intrusion, but as has been explained, this has hitherto enriched the vein. Therefore I see no reason why it should not continue to do so. $50,000 to $75,000 spent in further opening up the Queen vein, Yellowstone and Alexandra, would likely open up enough of pay ore to pay dividends on a fair capitalization for many years.

It will be noticed, I have made no attempt to estimate the quantity of ore in sight. Ore at the Queen has been so easily got that it has never been the policy to block out much ore, but to work out one chute before another was opened out. The Buckley Chute, which is at present being mined, will keep the mill running for some considerable time.

Attached hereto are "sketch" plans and maps.

Respectfully submitted,

Alexander Sharp

Mining Engineer.

2526, Seventh Avenue West,
Vancouver, B. C.

SHEEP CREEK CAMP

Dr. R. W. Brook, Director of the Geological Survey of Canada, reports:—
"A few days were spent examining the developments on Sheep Creek, near Salmo. A reconnaisance survey had of this district been made in 1897 by Mr. R. G. McConnell, assisted by the writer, who drew the attention of prospectors, who had been confining themselves to silver-lead and iron caps, to the promising quartz veins. A number of quartz claims were staked and a few developed with more or less success, notably the Yellowstone, on which a 10 stamp mill was erected, and the Queen, but only recently has any marked interest been taken in this camp. About $250,000 will be recovered from the limited operations of this year, and it gives promise of receiving vigorous development.

GEOLOGY

"The veins at present being worked occur in a band of quartzites, slates and schists, which extend northward from about Lost Mountains across Sheep Creek, at the forks of Sheep and Wolf Creeks, and up the ridge between Sheep Creek and Fawn Creek. To the west is a wide band of crystalline limestone. Some granitic and aplitic dykes are intruded into the formation and also some basic mica dikes. The general strike of the rocks is about N 12 degrees E with a dip of 50 degrees to the east. The veins are fissure veins cutting the formation, usually the quartzite.

QUEEN MINE

"The Queen vein on the south side of Wolf Creek, near the forks, is a well marked quartz vein from 6 to 11 feet in width, probably averaging about 7 feet in the workings. It is very regular and has usually clean-cut walls, with seams traversing it parallel to the walls.

"The country rock is white quartzite, with micaceous partings striking about north, and dipping eastward at an angle of 50 degrees. The vein, which is about vertical, strikes in a south-westerly direction, thus angling across the formation, but stringers from the vein may take off parallel to the quartzite. A few mica dykes cut the vein, the largest being about 25 feet wide. While the walls are usually clean-cut, in places the quartzite may be mineralized to some extent, in which case it is difficult to distinguish the vein matter from the white quartzite, except by the bedding planes of the latter, which differ in direction from the parting planes of the vein.

"The ore is white, milky quartz, with pyrite and pyrrhotite in about equal proportions developed in it. There is also some galena, and a sprinkling of blende and chalcopyrite. These sulphides constitute about 8% of the ore. The values are chiefly in gold, more than 50% of which is saved on the plates. The concentrates are said to run $48 to $70 per ton. The gold is reported to increase with the percentage of galena present suggesting a relationship between the two.

"The main tunnel has been driven 900 feet in on the vein. At 575 feet a 200 foot shaft has been sunk and a second level started from the shaft. All the workings are said to be in ore, which, in the bottom level, is reported to run higher than in the upper level. The ore as milled is reported to run from $10 to $18.

"The ore is treated in the Yellowstone mill, run by water power. About 55 tons of concentrates are produced. The Yellowstone lies between the forks of Sheep and Wolf Creeks, and was the banner claim, but work at present is confined to the Queen. I was informed by the manager that in all 30,000 tons of ore had been treated in the Yellowstone mill, producing about $370,000."

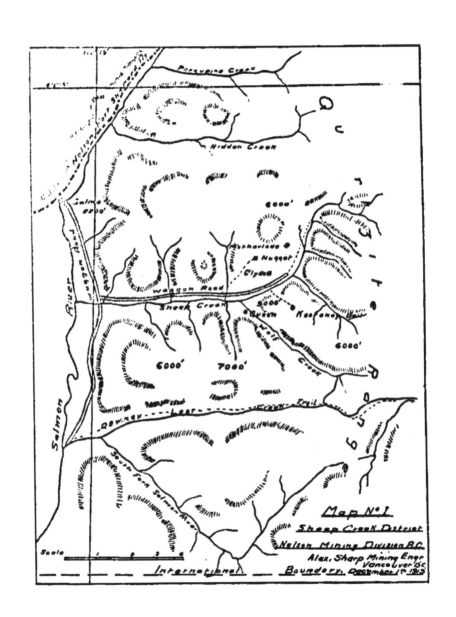

Map No. 1
Sheep Creek District
Nelson Mining Division B.C.
Alex. Sharp Mining Engr.
Vancouver B.C.
Boundary, December 1st 1913

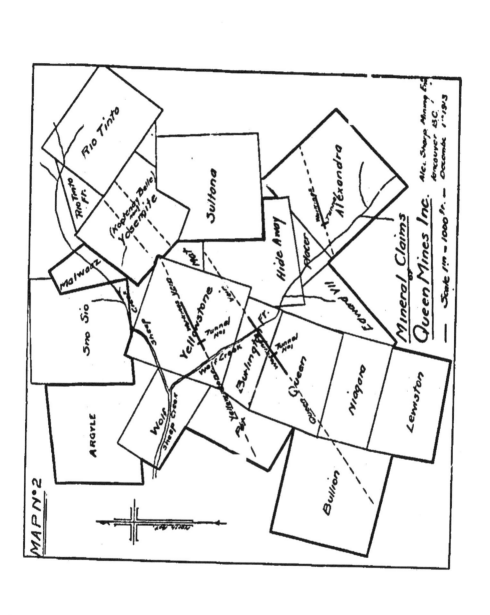

MAP N°2

Mineral Claims
of
Queen Mines Inc.
Alex. Sharp Mining Co.
Vancouver B.C.
December 1st 1913
— Scale 1m = 1000 ft. —

Rio Tinto

Rio Tinto Fr.

(Kootenay Belle) and
Yosemite

Malwazz

Sno Sio

ARGYLE

Wolf

Sheep Creek

Josie

Yellowstone

Tunnel No.1

Nat

Yellowstone

Yellowstone Fr.

Yellowstone Creek

Pat

Burlington Fr.

Tunnel No.2

Queen

Queen Fr.

Niagara

Lewiston

Bullion

Sultona

Hide Away

Placer

Edward

Alexandra

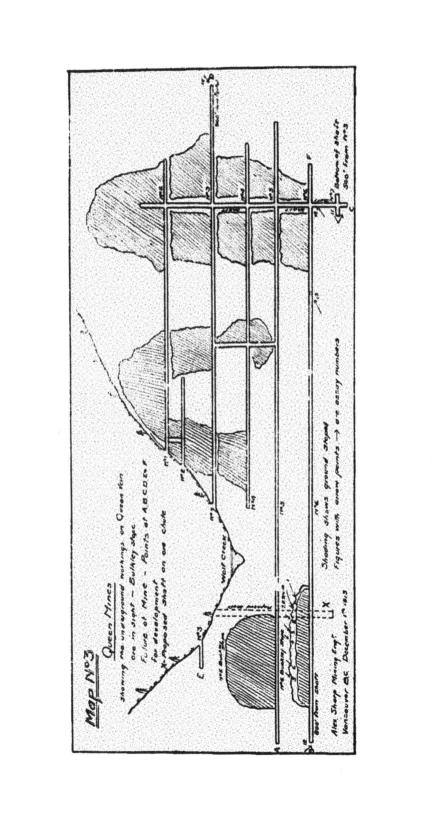

Map Nᵒ 3

Queen Mines

Showing the undeveloped workings on Queen Tom
ore in sight — Bulhig slope
F. Line of Mine — Points at A,B,C,D,E,F.
for development
X proposed shaft on ore chute

Wolf Creek

Showing shows ground striped
Figures with arrow points → are assay numbers

Bottom of Shaft
500' from Nᵒ 3

Alex Sharp Mining Eng't
Vancouver B.C. December 1ˢᵗ 1913